ERUPTION!

ERUPTION!

Mount Ruapehu Awakes

BRUCE HOUGHTON, VINCE NEALL and DAVID JOHNSTON

VIKING

The Authors

Bruce Houghton is a programme leader at the Taupo Volcanic Observatory within the Institute of Geological & Nuclear Sciences. He has worked extensively on eruption studies in New Zealand at White Island, Mayor Island, Auckland, Tarawera, Taupo, Ruapehu and Tongariro and at volcanoes in Alaska, Canary Islands, Greece, Hawaii, Italy and Germany. His research includes the first volcanic hazard assessment and the first interpretive model for the growth of Ruapehu volcano. He is a former leader of the International Commission on Explosive Volcanism and Deputy Secretary-General to the International Association of Volcanology and Chemistry of the Earth's Interior.

Vince Neall is an Associate Professor in the Department of Soil Science at Massey University. He is responsible for the Earth Science teaching programme and chairs the Civil Defence Board of Studies at Massey University. His major research interests have centred on the past volcanic histories of Egmont and Ruapehu volcanoes with a view to assessing future potential volcanic hazards. He has specialist expertise in lahars and tephras from volcanoes of this type. Currently he is convenor of the Ministry of Civil Defence's Volcanic Hazards Working Group, President of The Science Centre Inc. and a member of New Zealand's UNESCO Science Sub-commission.

David Johnston is a PhD student in the Department of Soil Science at Massey University, funded by a fellowship from the Institute of Geological & Nuclear Sciences. His research studies are physical and social impacts of volcanic eruptions — creating and testing eruption scenarios and assessing volcanic risk and the vulnerability to volcanic activity of lifelines in New Zealand communities.

Acknowledgements

The first acknowledgement is to the large groups of colleagues involved in the scientific response to the 1995-96 eruptions, namely Grey Arehart, Stewart Bennie, Hugh Bibby, Carol Bryan, Rod Burt, Grant Caldwell, Bruce Christenson, Shane Cronin, Sue Donoghue, Jackie Duncan, Brian Ferris, Jan Harris, Dan Hayba, Dianne Hewitt, Barbara Hobden, Katy Hodgson, Zinzuni Jurado-Chichay, Jérôme Lecointre, Jan Lindsay, Vernon Manville, Mitsuhiro Nakagawa, Ian Nairn, Peter Otway, Alan Palmer, Dinah Riley, Brian Robinson, Michael Rosenberg, Steve Sherburn, Brad Scott, Richard Smith, Bob Stewart, Thor Thordarson, Clel Wallace, Colin Wilson and Peter Wood.

We also gratefully acknowledge the contributions of staff from the Ministry of Civil Defence and the Volcanic Hazards Working Group, Department of Conservation, Ruapehu and Taupo district councils, Ruapehu Alpine Lifts, Turoa Ski Resort, Civil Aviation Authority, Ministry of Police, Television New Zealand, TV3, Environment Waikato, Tranz Rail and Contact Energy.

We also wish to thank the following individuals and companies for permission to reproduce their photographs in this book: Rod Burt, Catherine Heath, Brett Hooker, Gisborne Herald, Hawke's Bay Herald-Tribune, David Kerr, Landcare Research Ltd, NZ Herald.

Finally we wish to thank John Funnell, Keith McKenzie and Pete Masters for their skill in taking us places we needed to go.

Institute of GEOLOGICAL & NUCLEAR SCIENCES Limited

MASSEY UNIVERSITY

Contents

INTRODUCTION

The 1995-96 eruptions of Ruapehu captured the imagination of the New Zealand public and the publication of this book marks the first anniversary of the start of this dramatic sequence of events . . . which are far from over.

The first warning signs to the present activity were seen as early as January 1995, in small steam explosions during what seemed to be a typical cycle of heating of Crater Lake. These cycles, lasting typically 12 to 15 months, have occurred repeatedly from 1980, often unnoticed even by people close to the volcano. The 1995 cycle however culminated in much larger activity, on a scale not seen at Ruapehu since 1945.

The first significant eruption occurred on 29 June 1995 destroying some of the Institute of Geological and Nuclear Sciences' monitoring equipment on the shore of the mountain's Crater Lake. Larger eruptions followed on 18 and 20 September 1995, but public attention didn't focus on the mountain until a couple of days later. Explosions at 4:57 pm on 23 September produced an eruption plume 12 kilometres high, threw rocks to distances of 1.5 kilometres from the vent and generated lahars (volcanic mudflows) in three river valleys. Still larger peaks of activity followed on 25 September and 11 and 14 October 1995, and later on 17 June and 6-8 and 20 July 1996, sending ash over 300 kilometres from the mountain. The pattern of eruptions has changed with time as the crater lake was progressively removed during the explosions, and the activity moved from short, isolated, single explosions to sustained events lasting minutes to hours.

Eruptions create hazards which cannot be contained or confined. While the effects of Ruapehu's recent activity to date have been disruptive and not destructive for much of the sequence, the cost of that disruption is currently in excess of $50,000,000. Principal impacts have been the closure of national and international airports, truncation of the 1995 and 1996 ski seasons, and the loss of generating capacity within the Tongariro hydroelectric power scheme. An unusual feature of the Ruapehu eruptions has been the management of the crisis, without formal declaration of a state of civil emergency.

Eruptions of the size of the 1995-96 events leave few long-lasting scars and it is unlikely that this eruption will leave much trace on the Ruapehu landscape. The closest recent analogue, the 1945 eruption, has disappeared from public view and the public consciousness. No recognisable products of the 1945 eruption remain today on Ruapehu, and few of the affected communities have written records of the effects. It is this realisation that has prompted the authors to write this book.

BRUCE HOUGHTON
VINCE NEALL
DAVID JOHNSTON

UNDER THE VOLCANO

Mount Ruapehu is a cone volcano like Mount St Helens in the US. Large cone volcanoes are constructed over intervals of tens to hundreds of thousands of years — the oldest dated rocks at Ruapehu are 230,000 years old — and often consist of several smaller cones superimposed one on top of the other as shown in the sketch at left.

Cone volcanoes are built by numerous small to medium sized eruptions, both explosive eruptions and eruptions producing lava flows. Many of the smaller eruptions leave no recognisable trace on the volcano. In the last 1800 years alone, Ruapehu has had 18 eruptions of similar size to the 1995-96 event. Such events appear to occur on average once every 50-100 years.

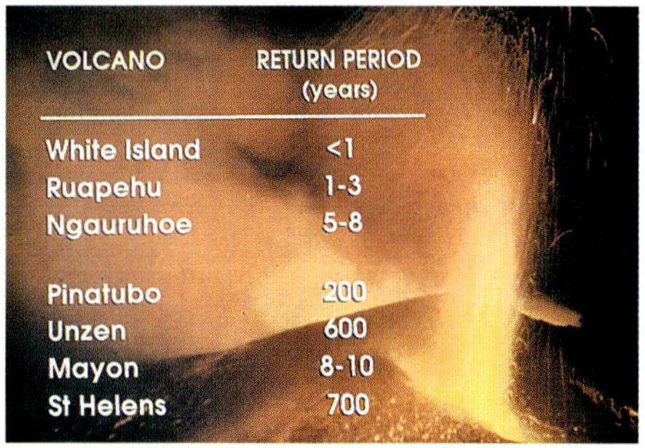

VOLCANO	RETURN PERIOD (years)
White Island	<1
Ruapehu	1-3
Ngauruhoe	5-8
Pinatubo	200
Unzen	600
Mayon	8-10
St Helens	700

Above: Like other New Zealand cone volcanoes, Ruapehu erupts very frequently by world standards. The table here compares the typical time-lapse between eruptions at Ruapehu with the behaviour of some of its overseas cousins.

Right: This drawing shows the complex plumbing system beneath the volcano. The new magma entering the plumbing system is torn apart in gas-driven explosions to generate the high plume of ash and gas above the volcano. Some of the material falling from the plume and landing on the steep, upper slopes of Ruapehu mixes with snow and ice to form a lahar.

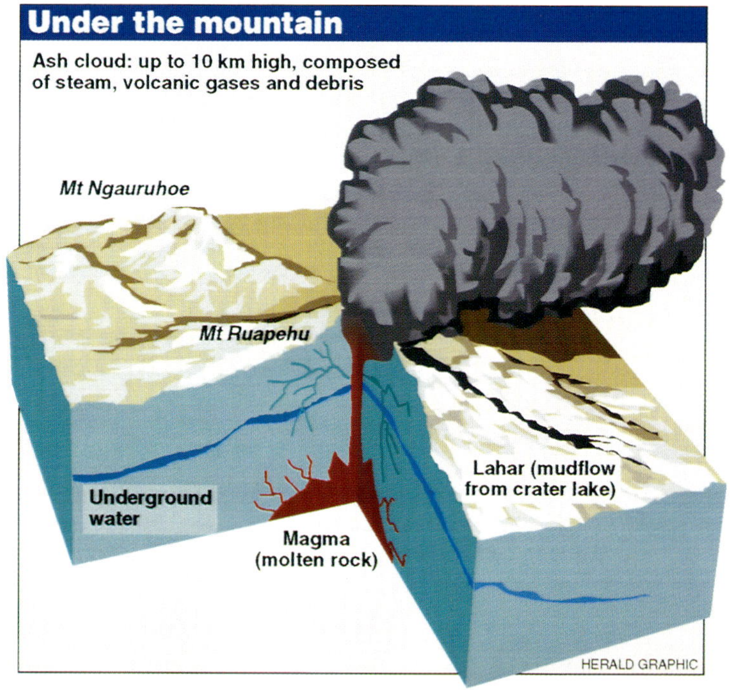

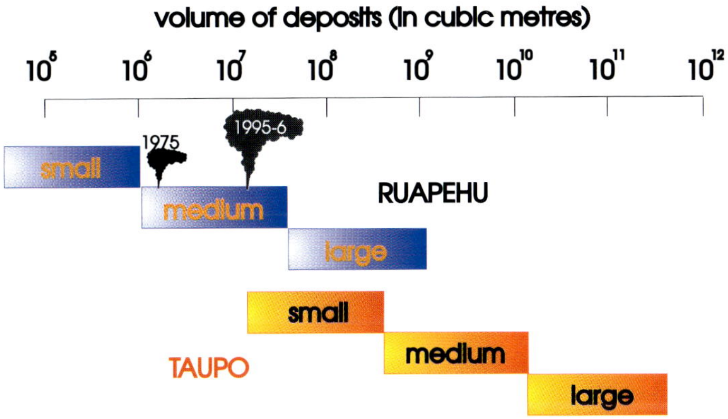

There is a major contrast between cone volcanoes like Ruapehu and caldera volcanoes like Taupo. The largest eruptions at Ruapehu are only slightly larger than the smallest events at Taupo, but Ruapehu erupts far more frequently.

We can place the 1995-96 eruptions on a size scale by comparing eruption plumes whose areas are proportional to the volume of erupted material. Here are shown Mt Ruapehu 1995-96 with Mt St Helens (1980), Tarawera (1886), Mt Pinatubo (1991), the Taupo eruption (181) and the largest eruption in the last 1000 years, that of Tambora in 1815. As will be seen, Ruapehu's 1995-96 eruptions have been small by world standards.

A unique feature of Ruapehu is the presence of the hot crater lake, set in a permanently glaciated environment. Eruptions at Ruapehu eject the hot lake water on to ice and snow creating frequent lahars.

Colin Wilson, GNS

MOUNT RUAPEHU AWAKES

The major activity in 1995 began at 8:07 am on 18 September, and went largely unnoticed by the public or the media. A moderate-sized explosion through Crater Lake produced a major mudflow (lahar) in the Whangaehu River destroying a foot bridge, and a smaller lahar on the Mangaturuturu Glacier. This event was accompanied by a magnitude 3.6 earthquake — the largest to occur in 1995.

Left: This photograph, taken on 19 September, shows the deposits of the lahar to the left side of the Whangaehu River. To the right, the river has already eroded away the rest of the deposit to re-establish its channel. A second moderate explosion took place at 1:22 am on 20 September generating a second lahar in the Whangaehu. Neither of these events impacted on the ski season which was in full cry by this time.

Right: New Zealand's attention became focused on Ruapehu at 5 pm on 23 September, five days after the first large eruption in 1995. As skiers packed up for the day, a sequence of three large explosions punched through Crater Lake, carrying lake water, ash and blocks torn from the magma (molten rock) below and the walls of the vent to heights of hundreds of metres above the summit of the mountain.

This photo captures the climax of the third and largest explosion. The image exploded around the world, attracting global attention to the 1995 eruptions. The eruption was accompanied by a magnitude 3.2 earthquake, shown in the insert sketch.

September

These three photographs capture the essence of the third and largest explosion on 23 September 1995, dramatically framed by the pylons of a Whakapapa ski lift.

<u>Below:</u> In the first image the black slug of water and ash has already separated into three jets. The column of white steam to the left is related to the two earlier explosions and is drifting peacefully away from the crater, divorced from the third explosion.

Catherine Heat

<u>Above:</u> In the second photograph, large blocks of rock and ice have detached from, and rise high above the black jets, drawing steam and water droplets behind them to form white vapour trails in their wakes. Some of the black jets were directed vertically, others travelled almost horizontally. Notice the spectacular low jet to the right which is arcing towards Paretetaitonga ridge.

Catherine Heath

Catherine Heath

Bruce Houghton, GNS

Above: In the final image, the third explosion has reached its peak. The blocks visible in the second photograph are now falling back on curving paths to intersect the ground surface. The largest of the black jets continues to rise vertically while the jet to the right is within seconds of smashing into the ridge of Paretetaitonga.

Top Right: This photograph, taken from a helicopter just 30 minutes after the third explosion, shows the dramatic effects of the eruption. The high peak in the photograph is Paretetaitonga, which is also visible in the other images. The 'fingerprint' of the spectacular inclined jet seen on the right-hand side of the explosion images is the black ash-stained area on the left of the photograph here. Spectacular impact craters pock-mark both the black ash and white snow. Each represents the impact point for one of the large blocks represented by the white vapour trails in the explosion photographs.

Explosions and Crater Lakes

When magma erupts through a lake the effect is similar to the explosions generated by pouring water into hot fat.

Heat from the 1100°C magma flashes some of the water into steam. The steam expands dramatically in volume, tearing the magma apart and flinging a mixture of steam, water, and rock particles into the air. This type of explosion is particularly unpredictable and difficult to anticipate.

This process occurred at Ruapehu in September and early October 1995, and on 17 June 1996 for as long as the crater lake remained.

Peter Otway, GNS

Left: A plume of steam and minor ash which rose to a height of 12 km was the final act of the 23 September eruption. This plume drifted to the north-east depositing a thin 'fingerprint' of ash. Notice how the base of the plume is no longer floating over the summit of Ruapehu but has been pushed sideways by the south-westerly wind and appears 'detached' from the volcano.

Opposite page: The summit of the mountain on 24 September showing the effects of the explosions of the day before. The wide, light-grey trail of ash on the left is the 'fingerprint' from the high eruption column shown opposite. To the right is the forked tongue of the Whakapapaiti lahar deposits. The black area at the summit is the region affected by the jets of ash, blocks and lake water. *Inset:* The summit of the mountain on 24 September is blackened by the deposits of the previous day's explosions. The grey steaming lake is reduced in size in comparison with before the eruptions.

September

Bruce Houghton, GNS

Left: During the explosions on 23 September, video cameras at Whakapapa recorded a lahar (volcanic mudflow) in the Whakapapaiti Valley on the north-western side of Ruapehu. This lahar is a mixture of ash, blocks of rock, Crater Lake water, and snow and ice which were collected along its path as it travelled at velocities of 15-20 metres per second. Other lahars passed down the Mangaturuturu and Whangaehu valleys on the western and eastern sides of Ruapehu.

Below: The lahar divided into two tongues to flow down valleys on either side of the Far West T-bar.

Bruce Houghton, GNS

Lahars

Lahars are flowing mixtures of water and volcanic debris generated during eruptions. Explosions through Crater Lake at Ruapehu eject large quantities of hot lake water onto the glaciers and snow fields. The hot water melts and mixes with snow to form the lahar which typically carries with it ash and blocks thrown out by the explosions. As lahars move down valleys they pick up and incorporate more snow and rock, drastically increasing the volume of the lahars. Other lahars at Ruapehu have formed when heavy rainfall affects thick unstable ash deposits high on the flanks of the volcano.

Right and below: The narrow right-hand portion of the lahar passed within metres of the control structure for the T-bar.

Bruce Houghton, GNS

Bruce Houghton, GNS

Hawke's Bay Herald-Tribune

Opposite page: The September eruption through Ruapehu's Crater Lake reached a peak on Monday 25 September between noon and 10 pm. Numerous small to moderate explosions through the lake, often only one to five minutes apart, began on the evening of 24 September. By the middle of the next day these explosions were feeding a continuous plume rising to 10 km above the mountain. Ash-fall from this plume closed the Desert Road for the first time during the eruption.

The explosions also created a pulsating series of lahars down the Whangaehu River, with a total volume equal to half the pre-eruption volume of Crater Lake.

Ian Nairn, GNS

Top right: This explosion was typical of the many events on 25 September.

Right: Blocks of rock ejected in the September explosions smashed through the walls of the Dome Shelter, only 300 metres from the lake. The shelter houses vital seismic equipment which has continued to function throughout the eruptions.

Ian Nairn, GNS

J. R. Grant

Top left: The lahar warning gauge on Whangaehu River early on the morning of 25 September just after the second lahar peak.

Right: As the lahars crossed the Whangaehu fan they spread into a series of anastomosing channels before all re-amalgamating further east where the main channel sharply turns southwards to flow along the Desert Road fault.

Beginning during the night of 24-25 September sustained expulsion of water from eruptions in the Crater Lake created a continuous multi-peaking lahar in the Whangaehu River for over 12 hours. During this time over 50 per cent of the Crater Lake was displaced. At daylight the second peak was passing Tangiwai with a maximum discharge of 160 cubic metres per second. The river had become a raging torrent which climbed 2.6 metres up the lahar warning gauge 10 kilometres upstream of Tangiwai. Below Tangiwai the lahar rose up to 3.8 metres, scouring vegetation from the banks and inundating quarries alongside the channel. By mid-morning a peak discharge of 230 cubic metres per second was recorded. A fourth peak was recorded in the late afternoon followed by a fifth smaller peak after dark.

Lloyd Homer, GNS

Vince Neall

Above: View of the top of the Whangaehu fan on 8 October. The large discharge lahars of 25, 27 and 29 September and 6 October filled the river channel from the cliff in distance to the foot of moraine in the foreground. Parts of the lahars overflowed to the north (by the white vehicle pictured) into the two most northern tributaries of the Whangaehu River. This portion of the lahar was within 500 metres of the Upper Waikato Stream catchment that flows into the Tongariro River.

Right: A 2.6 metre-sized boulder carried onto the Whangaehu fan by the 25 September lahars. In the distance steam clouds can be seen rising from Ruapehu's Crater Lake.

Vince Neall

Ian Nairn, GNS

The second peak to the 1995 activity occurred between 7 and 15 October. These explosions were subtly different from the September events. The events lasted much longer and were accompanied by sustained seismic activity.

The two major events occurred, firstly, between 9 pm on 11 October and 5 am on 12 October and, secondly, between 3 and 10 pm on 14 October. The first of these events caused the final emptying of Crater Lake. Both these eruptions produced high plumes which drifted at least 250 km downwind, raining ash over the countryside.

Left: A black ash-rich plume developing on 12 October, after the destruction of Crater Lake.

Ash fall

The jet of steam and gas formed during a sustained eruption can carry within it large quantities of ash and scoria. These rock particles rise high in the eruption plume and then drift downwind from the volcano and rain out gently to form a layer that mantles the landscape. The Ruapehu plumes drifted slowly away from the volcano so that distant communities recorded ash fall often hours after an eruption had ended. The ash fall becomes thinner and finer grained with distance from the volcano (see map at right). The eruption plume may temporarily obscure the sun creating an illusion of nightfall. Ash falls are generally not a hazard to life but create a host of minor problems and disruptions to everyday routine.

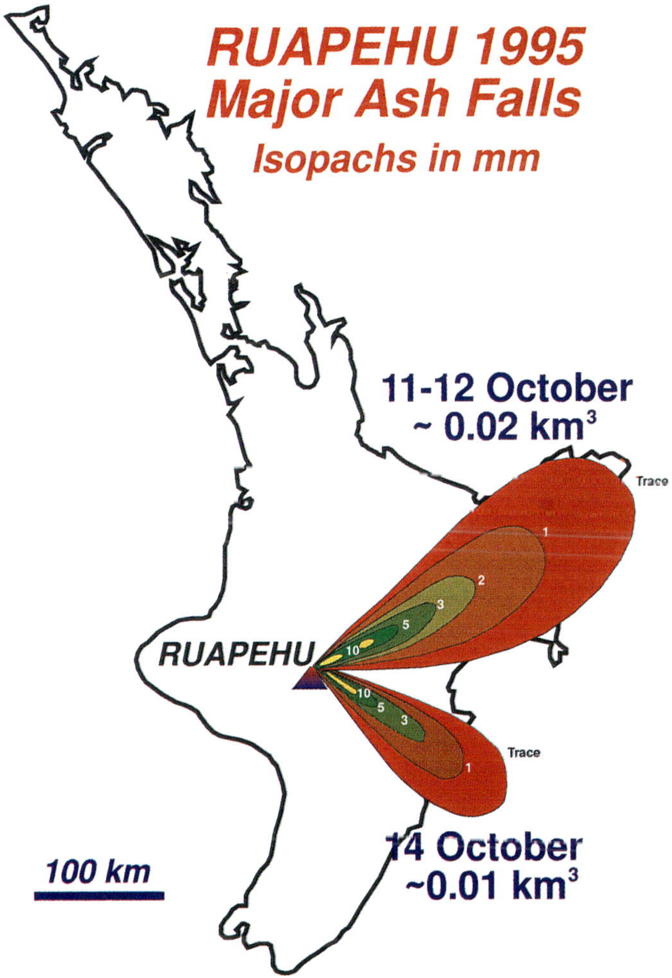

Thor Thordarson, GNS

Above: Rain-out of curtains of black ash from the 7 October eruption.

Left: A map prepared by the Institute of Geological and Nuclear Sciences showing the distribution of the ash falls on 11, 12 and 14 October. The map shows zones with different thicknesses of ash; the numbers show these thickness values in millimetres. Note how the distribution of the ash is strongly controlled by the prevailing wind direction — a south-westerly on 11 October and a north-westerly on 14 October. Communities north of Ruapehu had to wait until 17 June 1996 to get their first serious exposure to ash fall.

Below: These sketches show the amount of energy released by the large explosions in October 1995. Note the extended duration of the events, and that the 14 October event was shorter but more intense (or stronger) than its predecessor.

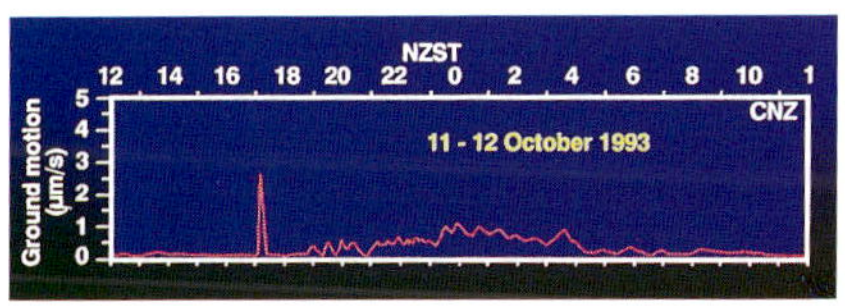

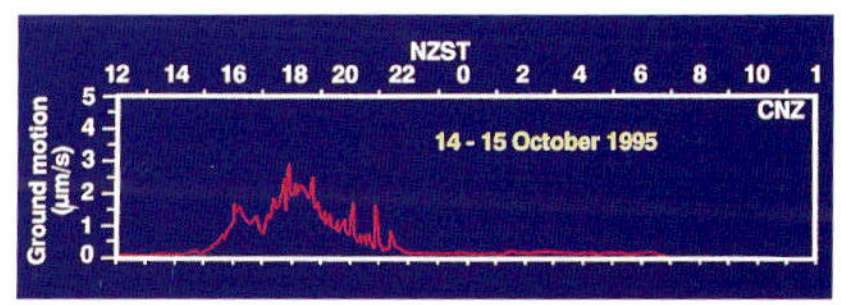

Magma and Gas

Molten rock (magma) typically contains between 1 and 4 per cent by weight of dissolved gas. As magma approaches the earth's surface the effect is somewhat similar to opening a shaken bottle of soft drink. The dissolved gas is released from the magma as bubbles which expand rapidly and tear the liquid magma apart to generate a spray of red-hot liquid droplets, or solid rock particles and gas.

Spectacular 'fire-fountaining' on the evening of 17 June 1996 tossed incandescent volcanic bombs and ash several hundred metres above the crater. These eruptions, although dramatic and accompanied by sonic booms and detonations, were not particularly energetic compared to the eruptions earlier that morning. This time-lapse image shows the glowing trails of the bombs as they arc through the air. New Zealand Herald

Gisborne Herald

New Zealand Herald

New Zealand Herald

Above: Heavy ash-falls associated with the eruptions caused problems for motorists and pedestrians alike.

Left: Some fish kills were reported in streams affected by lahars and by re-mobilisation of the ash into the waterways, but the death toll was fortunately low.

Far left: Ash is cleared from Wairoa streets.

Gisborne Herald

Greg Arehart, GNS

Top left: Sweatshirts are used as impromptu gasmasks to cope with the October eruptions.

Above: This desolate, ash-covered scene was the view at the normally bustling Tukino skifield on 3 October 1995.

Right: Local businesses rapidly adapted to the opportunities supplied by the 1995 eruptions.

David Johnston, GNS

David Johnston, GNS

The intensity of eruptions dropped gradually through late October and November 1995. By mid-November two small lakes had formed in the crater, and by late November they had merged.

Above: The summit of Ruapehu was a spectacular sight in March 1996 when this photograph was taken, with ash and blocks covering the crater rim and steam rising from around the newly-reformed lake. Note the widened crevasses on the right-hand side of the image. Visible in the foreground is the Dome Shelter which houses monitoring equipment.

Heavy rainstorms continued to wash ash into rivers long after the main eruptions in 1995 had ended, and new mudflows formed in some streams previously unaffected by the large eruptions. In late October 1995 ash began to enter tributaries of the Tongariro River — for the first time threatening the trout fishery at Taupo. A mudflow on 28 October destroyed a bridge 8 km downstream, and forced the closure of the Rangipo hydroelectric power station. Six other eruptions occurred between 31 October 1995 and 21 April 1996. Such problems are likely to persist, with diminishing effects, for years.

This sequence of three images shows the process from (1) the mountain (the headwaters of the Wahianoa Stream), to (2) the stream, to (3) the lake.

Bruce Houghton, GNS

Vernon Manville, GNS

October

Brad Scott, GNS

The last significant ash emission of 1995 occurred on 18 October although gas levels remained significant until March 1996. All the parameters that were monitored stayed at steady, moderate levels through the first half of 1996, and a decision to lower the alert level from 2 ('significant local eruption in progress') to 1 ('signs of unrest') was made on 11 June.

On 15 June seismic activity began a steady build-up at Ruapehu and the Institute of Geological and Nuclear Sciences issued a warning to the civil authorities. Around dawn on 17 June the first of several vigorous eruption plumes formed over the crater. Larger events during the next couple of hours threw blocks out of the lake and the latter formed a lahar in the Whangaehu River.

Continuing activity during the morning built-up a long-lived ash plume that was blown northwards by strong winds to cover Rotorua and Te Puke. Activity was recorded until early afternoon, then again from mid afternoon onwards, culminating in spectacular fire-fountaining eruptions between 9 and 10 pm. *(See pages 24 and 25.)*

Thor Thordarson, GNS

Right: By late morning the giant plume from the eruptions of 17 June was a spectacular sight when viewed from Taupo The majority of the ash missed Taupo, falling immediately to the west at Kinloch and Acacia Bay.

David Johnston, GNS

Bruce Houghton, GNS

Left: The vigorous explosions of 17 June sent convoluting pulses of grey ash and gas up to elevations of 6000 to 7000 metres.

Opposite page: Scientists and news crews monitored events using helicopters stationed upwind from the plumes.
Inset: Satellite cameras caught this dramatic image of the 17 June eruption plume as it drifted to an altitude of 20 km to the north-east of the volcano.

Below: The fall of ash from this eruption onto the snow pack spelt the first frustrating delay for skiers. But someone was keeping a happy face.

New Zealand Herald

June

David Johnston, GNS

David Johnston, GNS

Powerful explosions continued into July. These three photographs show the development of one pulse of eruption at midday on 7 July, growing larger and being progressively pushed sideways by the strong prevailing wind.

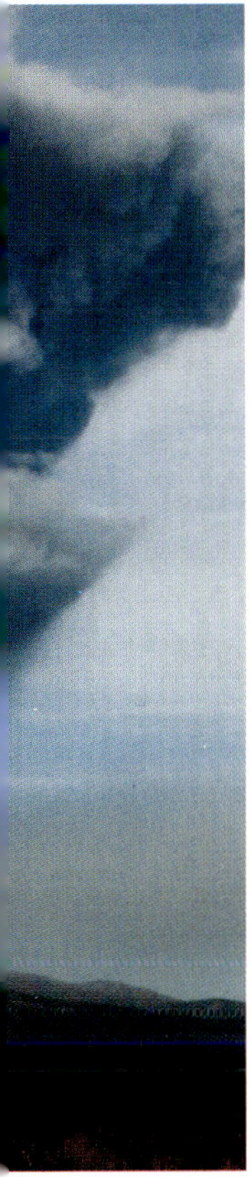

David Johnston, GNS

Bruce Houghton, GNS

July

Ash Fall

During strong explosions ash fall commences from even the lower portions of the plume. In this photograph, taken at 10:30 am on 10 July, a curtain of black ash descends from the left side of the plume. The distribution of ash is strongly affected by the wind — notice the clean snow in the foreground on the upwind side of the crater.

Opposite page: The July explosions ejected hot lava blocks and bombs as well as ash. The blocks and bombs formed impact craters in the snow when they landed, leaving the landscape pocked by craters. The hottest blocks then melted some of the underlying snow deepening the craters. The large black crater in this picture is 4 metres across. The blocks can push further upwind than the ash — so these blocks have landed on fresh snow on the wall of the crater which is not covered with ash.

Right: Early morning fire-fountaining was observed on 20 July, accompanied by sonic booms heard as far away as Turangi and Waikaremoana. Activity changed during the day, with explosions becoming stronger, and the highest eruption plumes rising to heights of 11,000 metres and drifting as far downwind as Opotiki. This image shows one of the major plumes at 4:06 pm, rising above cloud ceiling to reach a height of 9000 metres.

◆

Volcanoes are notorious for ignoring the advice of scientists, and Ruapehu has already surprised several times in 1995 and 1996. The most likely scenario for the near future is that activity will die away progressively, with highs and lows separated by intervals of several days to several weeks. In September 1996 Ruapehu continued to produce a high level of volcanic gas, implying that magma remains high in the throat of the volcano. While this situation remains, the volcano stays 'primed' with the potential to resume moderate sized eruptions relatively easily.

Bruce Houghton GNS

Bruce Houghton, GNS

THE CHANGING FACE OF RUAPEHU

As each new layer of ash forms on the summit of Ruapehu other processes are acting to transform and alter it. Rain storms carve and erode the ash into furrows, snow avalanches detach and carry the ash downhill, rocks slide and tumble down the steep slopes. The summit glaciers change in response to the eruption and the loss of the Crater Lake reveals landscapes inherited from the recent past. These images capture the changing face of the summit of this beautiful mountain in 1995 and 1996.

Left: The near-vertical cliff cut from the ice of the summit glaciers now disguised under ash and blocks. Large cracks are present in the ice curving parallel to the cliff edge.

Right: Here, on the outer slopes of the Pyramid, erosion by water ejected from the lake and by rainstorms has carved a pattern of steep, deep gullies across the once smooth landscape.

Below: Silver streaks washed by melt water run down across the glacier above the lake outlet — only to vanish abruptly into crevasses in the ice.

Thor Thordarson, GNS

Bruce Houghton, GNS

Lloyd Homer, GNS; (Inset) NZ Herald

THE VANISHING LAKE

The acidic Crater Lake is almost unique among the world's volcanoes as a hot lake set in a glaciated environment. The lake was destroyed by volcanic activity during the 1945 eruption, and slowly refilled between 1946 and 1953. When the lake is present, it strongly influences the style of Ruapehu eruptions: in particular the high frequency of lahars on the mountain is related to the mixing of erupted lake water with surface snow and ice.

Left, inset: Swimming in Crater Lake, once an important part of the recreational spectrum at Ruapehu, ceased after it was discovered that volcanic gases had made the lake strongly acidic.

Below: This diagram is a model for Crater Lake's behaviour prior to the 1995 eruptions, based on studies by Dr B. W. Christenson. The lake has two vents, with the major vent sitting over the conduit (or plumbing system) that was active in 1945 and in smaller later eruptions in 1969, 1971 and 1975. The temperature of the water increases steeply downwards, and a pool of molten sulphur lies on the bed of the lake.

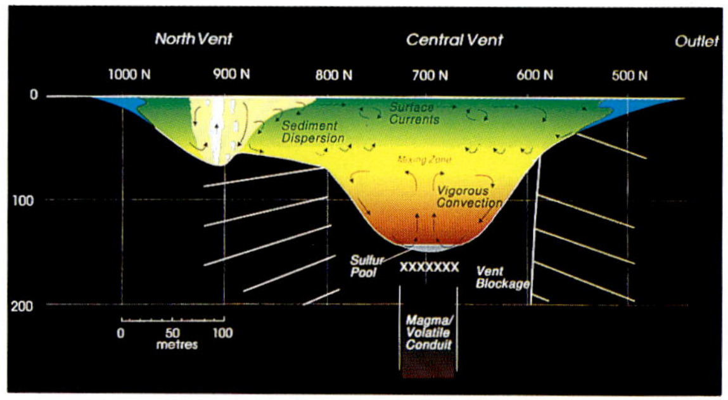

Bruce Christenson, GNS

These three photographs show the changing face of the Ruapehu crater in 1995.

Right: A Crater Lake greatly reduced in size is visible on 4 October 1995. The dark grey bench to the left of the shoreline had not been visible since the time of the 1945 Ruapehu eruption.

Below right: The lake vanished in the eruptions of 11-12 October leaving behind a broad, open crater with two vents clearly visible here on 18 October.

Below: Reduction of activity in late-October was followed in mid-November by the re-establishment of two small ponds on the floor of the crater. These ponds merged to form a new lake that was in turn destroyed on 17 June 1996.

Rod Burt, UC

Michael Rosen, GNS

Ian Nairn, GNS

THE SCIENCE OF VOLCANOES

Scientists use three techniques to monitor the 'pulse' of an eruption. These are: volcanic earthquakes (seismicity), ground deformation and gas studies.

Seismicity is the most useful tool because different types of seismicity can be associated with different processes beneath the surface — some types of activity are associated with the movement of molten rock or magma; others with the cracking of the walls of the vent and some with powerful explosions. The amount of energy being released at different times can also be compared.

At Ruapehu there is a permanent network of seismometers encircling the mountain.

Below: This chart is a portion of the seismograph record for Whakapapa village on 20 July. It contains numerous explosion earthquakes, each associated with a high ash plume. The photograph at right shows one of these plumes rising to an elevation of 9000 metres in the late afternoon of that day.

Bruce Houghton, GNS

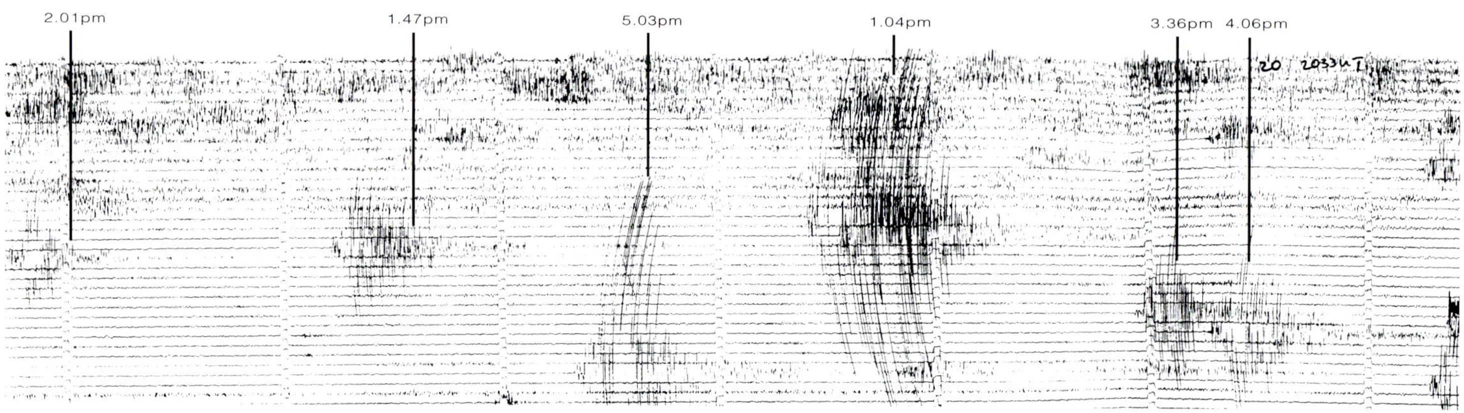

Bruce Houghton, GNS

Right: During the eruption seismologists installed additional temporary seismometers to gain more information about the activity. In particular, a cluster of extra stations were deployed further away from the crater to study the deeper parts of the 'plumbing system' beneath the volcano.

Below: The signals from all seismometers in the volcanic zone travel by fibre-optical cable to the volcano observatory at the Wairakei Research Centre. A duplicate of this laboratory has been established in Rotorua. The incoming signals pass through a system for automatic data analysis and are also recorded on the drums shown in this picture.

Bruce Houghton, GNS

Lloyd Homer, GNS

Sulphur Dioxide

The erupting magma carries with it, and releases, large quantities of gas — mostly steam but also sulphur dioxide (SO_2), hydrogen sulphide (H_2S), carbon dioxide (CO_2), and hydrogen chloride (HCl). Release of gas with only minor ash produces volcanic smog (vog). Vog results from chemical reactions between the SO_2, and oxygen, water and sunlight. These create droplets of acidic water and tiny particles (of sulphate minerals) which interfere with light rays from the sun, producing haze and smog. Vog results from a combination of volcanic and meteorological effects. Normally SO_2 is rapidly dissipated by wind at Ruapehu and diluted before these reactions have occurred. However in late October 1995 and late July 1996 a combination of light winds and high gas release produced spectacular effects. This photograph of July 1996 shows a pale gray haze of vog on the lower slopes of Ruapehu and Ngauruhoe (right) beneath a dramatic eruption cloud.

Bruce Houghton, GNS

Right: During 1995-1996, scientists supplied hazard maps as a management tool for the agencies involved with the civil response to the eruptions. These maps identified zones at risk from different types of hazards, for example lahars or falling blocks and bombs, and also set different levels of risk ie. red, orange, and yellow. This map shows the situation on 15 November 1995 at alert level 3. **Red1** indicates high risk from ballistic fall, pyroclastic fall/surge and volcanic gas; **Red 2**, high risk from lahar hazards; **Orange 1**, lesser ballistic, pyroclastic and gas hazards; **Orange 2**, lesser lahar hazards; **Yellow**, minor secondary lahar hazard.

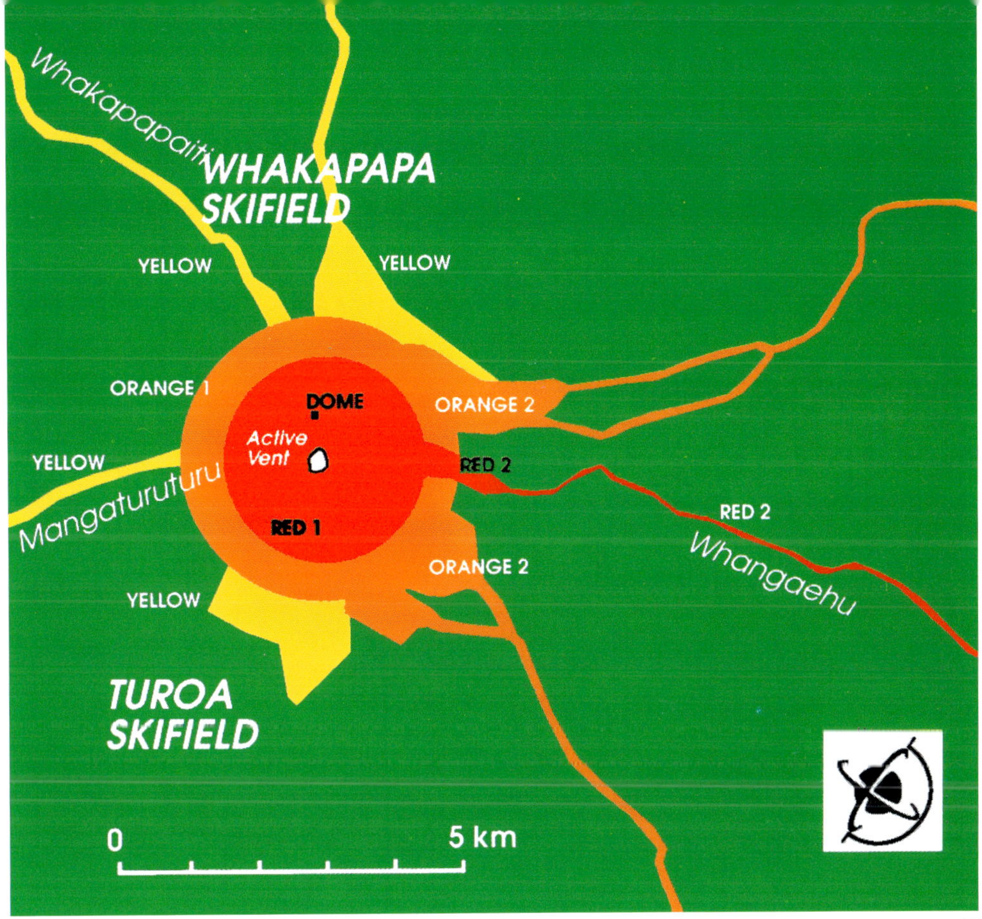

GNS

David Johnston, GNS

Left: A striking feature of the eruptions between September 1995 and August 1996 was that the civil response was managed without resorting to a formal declaration of a state of civil emergency. This was possible because (1) there was very effective liaison between the parties involved in the management exercise and (2) the highly hazardous zones were entirely within a single land unit (Tongariro National Park) with a strong, flexible management infrastructure. This book should be dedicated to the individuals who gave their time to the civil response to the Ruapehu eruption.

TIMELINE AND SUMMARY

KEY DATES

1995

January	Weak steam explosions during cycle of heating in Crater Lake.
29 June	Small explosion through lake destroys some monitoring gear.
18 and 20 September	Single moderate-sized explosions and lahars.
23 September	Single explosion throws blocks to 1.5 km distance, lahars in 3 valleys.
25 September	Numerous explosions, 10 km high plume, multiple lahars.
27 and 29 September	Explosions producing lahars.
7 October	Sustained 15-minute explosion column to 8-km height.
11-12 October	Largest sustained eruption in 1995-96. Ash to Gisborne.
11-12 October	**lake emptied**
14 October	Second largest sustained eruption. Ash fall to Hawkes Bay coast.
16-29 October	Strong gas emissions, declining ash output.
28 Oct-1 Nov	Rainfall remobilises ash as large lahars.
15 November	**two small lakes formed**
29 November	**two lakes have merged**

1996

15 June	Seismic tremor recommences.
17-18 June	Major plume to 20 km height. Ash to Waihi & Whakatane.
17 June	**lake emptied**
27 June	Plume to 6 km.
6-8,10 and15 July	Moderate eruptions, plumes to 6 km, ash fall on adjacent communities.
20 July	Plume to 10.5 km height, blocks fall to 1.4 km, ash fall at Taupo.

KEY FACTS

The 1995-96 events:

- Comprised the largest and mostly costly eruption in New Zealand since Mt Tarawera in 1886.
- Produced 0.06 km^3 of new material over an area of more than 30,000 km^2.
- Deposited ash more than 300 km from its source.
- Created more lahars (36 in Whangaehu Valley alone by 14 August 1996) than the total for all previous eruptions this century.

KEY CONSEQUENCES

The events:

- Closed 11 airports from Auckland to Wellington
- Caused approximately $9.5 million in damage to hydroelectric power facilities.
- Closed State Highway 1 on at least three occasions.
- Killed deer, trout in two rivers, and 2000 sheep in the areas affected by ash fall.
- Closed three Ruapehu skifields in 1995 and 1996.
- Enhanced soil sulphate levels to up to 25 times pre-eruption levels.

VIKING
Penguin Books (NZ) Ltd, 182-190 Wairau Road, Auckland 10, New Zealand
Penguin Books Ltd, 27 Wrights Lane, London W8 5TZ, England
Penguin USA, 375 Hudson Street, New York, NY 10014, United States
Penguin Books Australia Ltd, 487 Maroondah Highway, Ringwood, Australia 3134
Penguin Books Canada Ltd, 10 Alcorn Avenue, Toronto, Ontario, Canada M4V 3B2

Penguin Books Ltd, Registered Offices: Harmondsworth, Middlesex, England

First published by Penguin Books (NZ) Ltd 1996
1 3 5 7 9 10 8 6 4 2

Front cover photograph by David Kerr/Greatscapes; back cover photographs *NZ Herald*; photographs pages 1, 2 and 6, *NZ Herald*; page 5, Keylight Photo Library.

Typesetting and text design Bona Fide Design Ltd, Auckland
Printed by Wyatt & Wilson Print Ltd, Christchurch
ISBN 0-670-87441-8